Flick a Switch

by Anna Prokos

BLACKBIRCH®
PRESS

San Diego • Detroit • New York • San Francisco • Cleveland • New Haven, Conn. • Waterville, Maine • London • Munich

For more information, contact
The Gale Group, Inc.
27500 Drake Rd.
Farmington Hills, MI 48331-3535
Or you can visit our Internet site at http://www.gale.com

LIBRARY OF CONGRESS CATALOGING-IN-PUBLICATION DATA

Prokos, Anna.
  Flick a switch / by Anna Prokos.
     p. cm. — (Step back science series)
Includes index.
Summary: Traces what happens when a light switch is turned on back through how electricity is generated, distributed, and installed in a house.
  ISBN 1-56711-676-0 (hardback : alk. paper)
  1. Electric power—Juvenile literature. [1. Electric power.] I. Title. II. Series.

TK148 .P76 2003
621.3—dc21   2002013161

Printed in United States
10 9 8 7 6 5 4 3 2 1

Contents

Flick a Switch

How to Use This Book

Each Step Back Science book traces the path of a science-based act backwards, from its result to its beginning.

Each double-page spread like the ones below explains one step in the process.

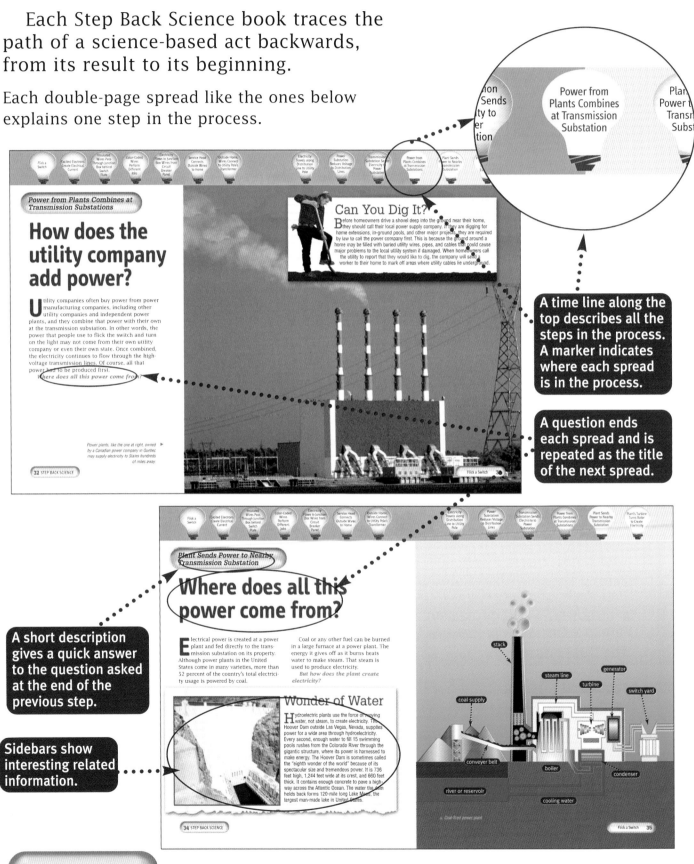

A time line along the top describes all the steps in the process. A marker indicates where each spread is in the process.

A question ends each spread and is repeated as the title of the next spread.

A short description gives a quick answer to the question asked at the end of the previous step.

Sidebars show interesting related information.

Power from Plants Combines at Transmission Substations

How does the utility company add power?

Utility companies often buy power from power manufacturing companies, including other utility companies and independent power plants, and they combine that power with their own at the transmission substation. In other words, the power that people use to flick the switch and turn on the light may not come from their own utility company or even their own state. Once combined, the electricity continues to flow through the high-voltage transmission lines. Of course, all that power had to be produced first.
Where does all this power come from?

Power plants, like the one at right, owned by a Canadian power company in Québec, may supply electricity to States hundreds of miles away.

32 STEP BACK SCIENCE

Can You Dig It?

Before homeowners drive a shovel deep into the ground near their home, they should call their local power supply company. If they are digging for home extensions, in-ground pools, and other major projects, they are required by law to call the power company first. This is because the ground around a home may be filled with buried utility wires, pipes, and cables that could cause major problems to the local utility system if damaged. When homeowners call the utility to report that they would like to dig, the company will send a worker to their home to mark off areas where utility cables lie underground.

Plant Sends Power to Nearby Transmission Substation

Where does all this power come from?

Electrical power is created at a power plant and fed directly to the transmission substation on its property. Although power plants in the United States come in many varieties, more than 52 percent of the country's total electricity usage is powered by coal.

Coal or any other fuel can be burned in a large furnace at a power plant. The energy it gives off as it burns heats water to make steam. That steam is used to produce electricity.
But how does the plant create electricity?

Wonder of Water

Hydroelectric plants use the force of moving water, not steam, to create electricity. The Hoover Dam outside Las Vegas, Nevada, supplies power for a wide area through hydroelectricity. Every second, enough water to fill 15 swimming pools rushes from the Colorado River through the gigantic structure, where its power is harnessed to make energy. The Hoover Dam is sometimes called the "eighth wonder of the world" because of its spectacular size and tremendous power. It is 736 feet high, 1,244 feet wide at its crest, and 660 feet thick. It contains enough concrete to pave a highway across the Atlantic Ocean. The water the dam holds back forms 120-mile long Lake Mead, the largest man-made lake in United States.

34 STEP BACK SCIENCE

A Coal-fired power plant

Flick a Switch 35

labels: stack, steam line, generator, turbine, switch yard, coal supply, conveyer belt, boiler, condenser, river or reservoir, cooling water

Side Step spreads, like the one below, offer separate but related information.

Every Side Step spread contains a sidebar.

The Big Picture, on pages 40–41, shows you the entire process at a glance.

Flick a
Switch

Excited Electrons
Create Electrical
Current

Insulated
Wires Pass
Through Junction
Box Behind
Switch
Plate

Color-Coded
Wires
Perform
Different
Jobs

Electricity
Flows to Junction
Box Wires from
Circuit
Breaker
Panel

Service Head
Connects
Outside Wires
to Home

Outside Home,
Wires Connect
to Utility Pole's
Transformer

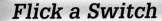

Flick a Switch

How does electricity light up a room?

Walk into a dark room and flick a switch to turn on the light. Sounds simple, but there is a lot going on behind the scenes. When flipped on, a switch allows electrical current to flow behind the switch plate and light up a room. Beyond this room, dozens of high-powered machines and thousands of miles of wires carry the electricity to power the lights—and all the other electrical devices—in a home.

What happens behind a light switch when it is flicked?

 Electricity Travels Along Distribution Line to Utility Pole

 Power Substation Reduces Voltage to Distribution Lines

 Transmission Substation Sends Electricity to Power Substation

 Power from Plants Combines at Transmission Substation

 Plant Sends Power to Nearby Transmission Substation

 Plant's Turbine Turns Rotor to Create Electricity

Excited Electrons Create Electrical Current

What happens behind a light switch when it is flicked?

When a light switch is flipped on, electricity flows through wires behind the switch plate to the light bulb and back again in a loop, or circuit. Charged particles called electrons rush through the wiring. They push each other forward in a line, and they move so fast that they create extreme heat—more than 4,000 degrees Fahrenheit (2,204 degrees Celsius). When the current reaches the lamp, this extreme heat is emitted as light. The process takes less than a second.

With all that heat, how do wires behind the switch plate carry electricity safely?

Back in Time

Before the use of electricity was widespread, people created light, heat, and refrigeration in various ways. In the 1700s and 1800s, people burned natural materials such as coal, petroleum, and wood to create light and heat. Foods were cooked over wood-burning pits or stoves, which also provided heat. In the winter, food was stored outdoors where the cold temperatures could keep it from spoiling. In warmer months, people placed food in an icebox, an airtight box kept cold by a huge chunk of ice. When the ice melted, another chunk would take its place. Until the mid-1800s, people used kerosene oil made from petroleum for light.

◄ *Oil-burning lamps were often used for lighting before the use of electricity.*

Electricity Travels Along Distribution Line to Utility Pole

Power Substation Reduces Voltage to Distribution Lines

Transmission Substation Sends Electricity to Power Substation

Power from Plants Combines at Transmission Substation

Plant Sends Power to Nearby Transmission Substation

 Plant's Turbine Turns Rotor to Create Electricity

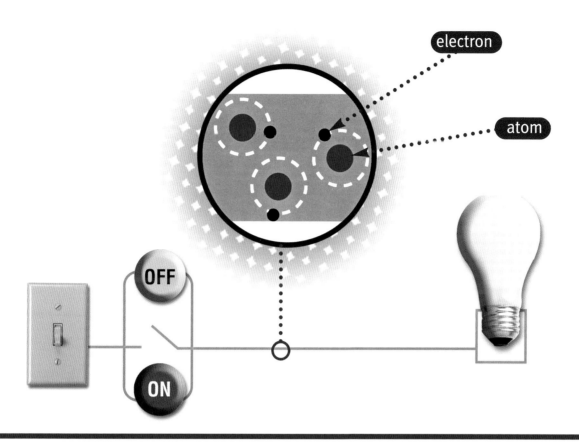

electron

atom

OFF

ON

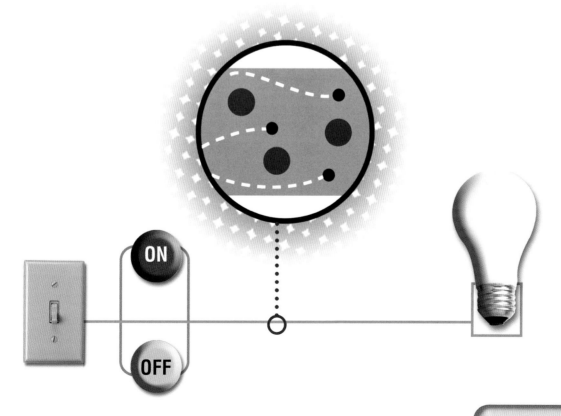

ON

OFF

Insulated Wires Pass Through Junction Box Behind Switch Plate

How do wires behind the switch plate carry electricity safely?

For safety, the metal wires behind the switch plate are covered with different-colored plastic or rubber insulation. Plastic and rubber are poor conductors of electricity, so they can safely contain the super-hot current.

Additionally, the wires are contained in a plastic or metal junction box. The junction box plays an important safety role in a home's electrical wiring. If a charged wire's protective coating has worn away to expose bare metal at its end, the current can take a detour from its proper circuit and hop off. Without a junction box to contain this runaway current, it would escape into the walls and could cause a fire.

How do the wires in the junction box all work together?

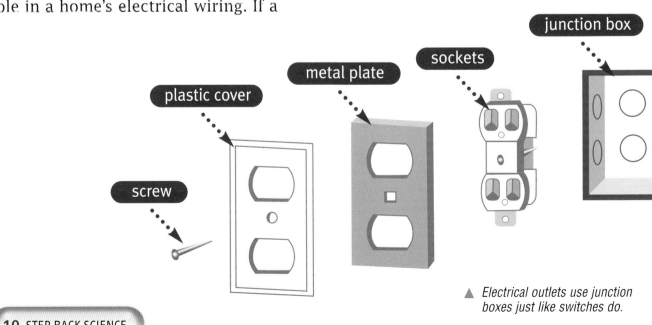

screw | plastic cover | metal plate | sockets | junction box

▲ *Electrical outlets use junction boxes just like switches do.*

junction box

switch

Not Just for Pennies

All metals are good conductors of electricity, but copper's atomic structure makes it one of the best. A copper atom has a single electron on its outermost shell. This lone electron is easily knocked out of position. When an electron leaves its atom, it jumps to a nearby atom, bumping another electron loose. This chain of jumping electrons creates the flow called current.

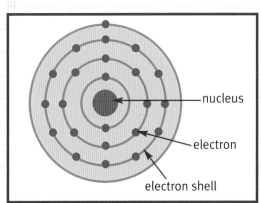

nucleus

electron

electron shell

▲ *Copper wire is an excellent conductor of electricity.*

Flick a Switch	Excited Electrons Create Electrical Current	Insulated Wires Pass Through Junction Box Behind Switch Plate	Color-Coded Wires Perform Different Jobs	Electricity Flows to Junction Box Wires from Circuit Breaker Panel	Service Head Connects Outside Wires to Home	Outside Home, Wires Connect to Utility Pole's Transformer

Color-Coded Wires Perform Different Jobs

How do the wires in the junction box all work together?

Each wire performs a different job, which is signified by the color of its protective coating: white, green, and red or black. Red or black wires are called hot wires, and they are charged with electricity. The hot wire carries the current to the light bulb. The white "neutral" wire, which is uncharged, carries the current back from the light bulb—when wiring is working properly. Sometimes a hot wire overloads, however, when the current cannot complete its intended circuit. For instance, worn-away plastic covering may leave a wire bare and allow it to touch the metal in the light fixture or the junction box, creating a large current and the risk of fire. This is called a short circuit. This is where the green or "grounding" wire comes in. It carries the "wrong way" electricity away from the short and out of the house.

But where does the electrical current in the junction box flow from?

Major Shock

Listen to that age-old warning: Never stick anything that isn't a plug into an electrical outlet. Most electrical outlets have a constant supply of electricity running through them. All it takes to activate that electricity is to plug something into the outlet.

Anything, not just a plug, can complete the circuit electricity is always trying to make. When the object makes contact with the wires inside the outlet, electricity surges through it. The jolt of electricity is so powerful that it can deliver a shock strong enough to send the person holding the object flying backward. It may also cause death by electrocution.

▲ *Electrical outlets can be dangerous.*

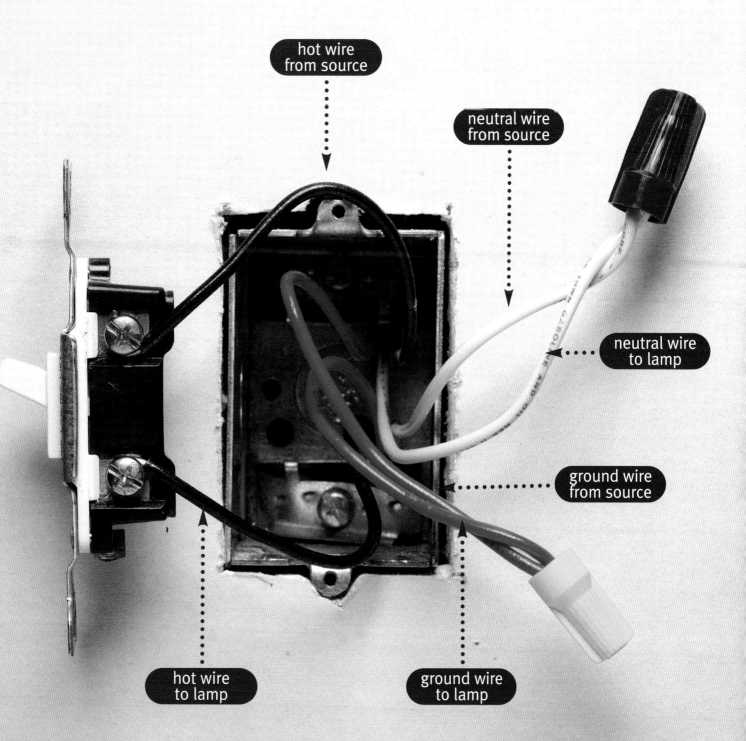

Electricity Travels Along Distribution Line to Utility Pole

Power Substation Reduces Voltage to Distribution Lines

Transmission Substation Sends Electricity to Power Substation

Power from Plants Combines at Transmission Substation

Plant Sends Power to Nearby Transmission Substation

Plant's Turbine Turns Rotor to Create Electricity

hot wire from source

neutral wire from source

neutral wire to lamp

ground wire from source

hot wire to lamp

ground wire to lamp

▲ The hot wire carries electricity to the light. The neutral and ground wires carry electricity away from the light and drain it safely into the ground.

Flick a Switch   **13**

Flick a Switch

Excited Electrons Create Electrical Current

Insulated Wires Pass Through Junction Box Behind Switch Plate

Color-Coded Wires Perform Different Jobs

Electricity Flows to Junction Box Wires from Circuit Breaker Panel

Service Head Connects Outside Wires to Home

Outside Home, Wires Connect to Utility Pole's Transformer

Electricity Flows to Junction Box Wires from Circuit Breaker Panel

Where does the electrical current in the junction box flow from?

The current flows from the home's main service panel, called a circuit breaker panel. This large, gray, rectangular box is often located in the basement, garage, or utility area. It has two rows of short switches, which are usually kept in the "on" position. Each switch controls one circuit.

Household current flows under pressure from the circuit breaker panel, around the house along hot wires, through the junction box, and to the switch. The pressure of the current is called voltage. The more powerful the appliance, the more voltage, or force, is necessary to make it work. If a circuit's voltage gets dangerously high, as when a short circuit occurs, the circuit breaker shuts the circuit off.

But how does the electricity get to the circuit breaker panel?

Easy to Con-Fuse

Some houses, particularly those built between 1950 and 1965, use fuses instead of circuit breakers. The two work a little differently, but both protect the electrical system from getting out of control.

A circuit breaker has a metal strip inside that heats up and bends when current passes through. If there is too much current, the strip bends and pulls the switch to the "off" position.

Inside a fuse's glass cap is a metal ribbon through which electricity flows to the service panel. If there is too much current, the metal ribbon melts and blocks the electrical flow. This is called blowing a fuse.

When a fuse blows or a circuit breaker needs to be reset, it can mean that too many things are taking power through that circuit.

Electricity
Travels Along
Distribution
Line to Utility
Pole

Power
Substation
Reduces Voltage
to Distribution
Lines

Transmission
Substation Sends
Electricity to
Power
Substation

Power from
Plants Combines
at Transmission
Substation

Plant Sends
Power to Nearby
Transmission
Substation

Plant's Turbine
Turns Rotor
to Create
Electricity

◀ *Circuit breakers
prevent fires and
save lives.*

Service Head Connects Outside Wires to Home

How does electricity get to the circuit breaker panel?

Electricity flows to the panel from the service head, located near the roof outside the home. This cane-shaped device holds a set of service wires from the local utility, or power supply, company and provides a point of entry into the home.

Typically, there are three wires hooked up to the service head. Two of the wires are insulated and thick enough to deliver large quantities of electricity. The third wire is bare. This is the grounding wire, which is responsible for carrying electricity to the ground.

The two insulated wires each carry 120 volts of electricity. Those wires allow the use of 120-volt and 240-volt appliances in a house. Both of the wires are combined at the service panel to power those appliances. Most of the gadgets and appliances in the home, such as a microwave or a computer, use 120-volt electricity. Only heavy-duty machines, like air conditioners and ovens, use 240-volt electricity.

But where do the power lines that connect to the service head come from?

Electricity enters the home at the service head. ▶

Electricity Travels Along Distribution Line to Utility Pole

Power Substation Reduces Voltage to Distribution Lines

Transmission Substation Sends Electricity to Power Substation

Power from Plants Combines at Transmission Substation

Plant Sends Power to Nearby Transmission Substation

Plant's Turbine Turns Rotor to Create Electricity

Batteries Not Included

Small electrical devices usually run on much lower voltage than a home's wiring carries. That is why they can run on batteries instead of being plugged in.

Batteries contain chemicals that react with each other to produce electrons. These are called electrochemical reactions. The electrons start to flow when a battery's negative terminal (-) is connected to its positive terminal (+) by a wire or metal strip. The kind of current batteries supply is called direct current. It means that the electrons flow one way in a continuous movement. Another example of direct current is static electricity, which can cause clothes to stick together when removed from a dryer.

The first battery was a stack of zinc, silver, and paper soaked with salt water. It was invented by Italian scientist Alessandro Volta in 1800, and called the voltaic pile.

ELECTRICITY GOING INTO HOME IS MONITORED

What keeps track of how much electricity flows inside the home?

The local utility company keeps track of electricity usage with a watt-hour meter. As electricity enters a home, it passes through the meter, usually located on the side of a home or in the basement, garage, or utility room.

A watt-hour meter measures the number of watts per hour of electricity used. Tiny dials on the meter inch ahead as more and more electricity is used. Most watt-hour meters have a tiny disk that spins around. This thin disc rotates as power is used.

The utility company usually sends someone to read the meter so the company knows how much to charge each homeowner for energy use each month. Reading an electric meter can be tricky, especially because it takes four or five separate dials to show how much energy is being used. Follow the steps at right to read the electric meter pictured above them.

▲ *Utility companies track power usage with watt-hour meters.*

Measuring Up

Measuring electricity is important to homeowners as well as to the utility company. If consumers understand how electricity is measured, they can make educated decisions about usage in their home. Brush up on these measuring terms:

Amps is the term for the electrical current that flows through a wire. A 100-watt light bulb draws about one amp of current.

Voltage is the pressure of electricity. Electrical current in a wire is driven by voltage. Most lights and appliances, such as microwaves, use 120 volts. Bigger appliances, such as air conditioners, use 240 volts.

Watts measure the amount of electrical power. To get the number of watts, multiply amps by voltage. A typical light bulb needs about 60 watts, while a central air conditioner requires 5,000 watts. One thousand watts equals one kilowatt.

Kilowatt-hour is what utility companies use to measure the electricity consumed in your home. Running a 100-watt light bulb for 10 hours uses one kilowatt-hour of electricity. An average home uses about 24 kilowatt-hours each day.

❶ Locate the dials; read them from left to right.

❷ Each dial has a pointer that points to a number. If the pointer falls between two numbers, the lower number is recorded. (In the picture above, the pointers all point to zero because the meter has not been used.)

❸ The dials show how many kilowatt-hours of electricity are used in a home. A kilowatt-hour is the total amount of energy used.

❹ To find out how much electricity was used in one month, subtract the previous month's reading from the current month's reading. The difference is the number of kilowatt-hours used in that month.

Flick a Switch

Excited Electrons Create Electrical Current

Insulated Wires Pass Through Junction Box Behind Switch Plate

Color-Coded Wires Perform Different Jobs

Electricity Flows to Junction Box Wires from Circuit Breaker Panel

Service Head Connects Outside Wires to Home

Outside Home, Wires Connect to Utility Pole's Transformer

Outside Home, Wires Connect to Utility Pole's Transformer

Where do the power lines that connect to the service head come from?

Those wires begin at utility poles placed on neighborhood streets. These poles commonly provide wires for telephones and cable TV as well as electricity. In some newer communities, power lines can be located underground.

At the top of each pole is a drum that delivers the right amount of electricity to homes. Called a transformer, this device transforms power so that a manageable amount of voltage enters the home. Normally, electricity travels through power lines at about 10,000 volts. The transformer must change all that power to 120 volts so it can be safely used in homes.

In addition to the wires that lead to the service head, every utility pole has a wire that leads down the pole to the ground and then 6 to 10 feet (2–3 m) under the surface. It carries any extra electricity safely below the earth.

How does the electricity get to the pole and transformer drum?

▲ *A group of brand-new transformers awaits use by a power company.*

Electricity Travels Along Distribution Line to Utility Pole

Power Substation Reduces Voltage to Distribution Lines

Transmission Substation Sends Electricity to Power Substation

Power from Plants Combines at Transmission Substation

Plant Sends Power to Nearby Transmission Substation

Plant's Turbine Turns Rotor to Create Electricity

Transformers like the ones on the pole above convert high-voltage electricity into a lower voltage safe and appropriate for household use.

Struck by Lightning

Utility poles are magnets for lightning. Because lightning commonly strikes tall objects, one bolt can knock out an entire electrical system. That is why many power lines and utility poles have a safety feature: a lightning wire or rod.

A lightning rod is a metal rod that connects to a thick copper or aluminum wire. This metal wire is connected to an electrical grid buried in the ground. Despite its name, a lightning rod does not attract lightning. Instead, it provides lightning with an easy path to the ground—away from high voltage wires and other structures.

◄ Lightning rods protect buildings and utility poles by giving lightning a quick route to the ground.

POWER'S OUT!

What happens if the power supply is cut off as it travels to a home?

A blackout can occur when there is not enough power to supply an area. As a result, the area receives no power at all. The appliances and devices that depend on electricity, such as lights, refrigerators, TVs, and computers, all stop working.

Although the local utility company works hard to keep up with electrical demands, sometimes the need for electricity is greater than the amount of electricity produced. In hot summer months, many people run their air conditioner at full power for 24 hours a day. The air conditioners used by an entire neighborhood, plus the other appliances they use each day, can create a huge demand for electricity.

Utility companies expect this summer increase in power demand, so they keep power plants working hard to make more electricity. Sometimes they cannot prevent a blackout, however, because the available electricity is limited to how much energy their plants can make.

When prolonged power shortages are expected, utility companies and

state energy groups issue alerts to residents. These alerts remind people of the high demand for electricity, and its limited supply. In emergencies, utility companies may even control energy overuse by issuing a rolling blackout for a particular area, which cuts off the flow of electricity to different homes at different times.

When a rolling blackout or other power problem occurs, some home appliances can be damaged. Many appliances, such as microwave ovens, cordless telephones, and home security systems, have tiny switches inside that provide information about how the appliance should run. When there is a change in the power supply, those switches can melt. To keep appliances safe, many homeowners install surge protectors. These devices protect appliances from overheating if there is a surge, or rush, of power when electric service is restored.

◀ A surge protector minimizes the risk of damage or fire.

The Nights the Lights Went Out

November 9, 1965, was a dark day in the history of the Northeast. Between 5:27 and 5:40 p.m., 80,000 square miles of the northeastern United States and Ontario, Canada, lost power. More than thirty million homes were left in the dark when a high-voltage transmission line disconnected from the main electric grid.

New York City was the hardest hit by the blackout. The city relies on electricity to power everything from underground subways to railroad tracks to the famous neon lights of Times Square. The blackout hit at the peak of rush hour. It took 14 hours to restore electricity to New York City. In the meantime, ten thousand people were stuck in subway cars that would not open for hours. Other people were stranded in elevators and train stations, unable to get home or even find a hotel in a crowded city filled with tourists and tired workers looking for a place to wait out the blackout.

▲ Bright lights usually shine in Times Square, New York City.

In 1977, New York City had another power outage that made headlines. This time, the cause was powerful lightning. Four successive lightning strikes to the north of the city knocked out major power lines. Although the city was connected to the transmission grids in New Jersey and Long Island, the power outage cut the power circuits that fed into the city. Some parts of the city remained in total darkness for up to 25 hours.

WEATHER WATCH

How does weather affect electricity?

About 70 percent of power outages are caused by weather conditions: lightning strikes, hurricanes, tornadoes, blizzards, and high winds. Because most of the electrical system is above ground, it is subject to nature's harsh elements.

Lightning can strike transformers, causing an abnormal increase in electrical voltage that can knock out power and even destroy parts of the electrical system. Lightning can also cause trouble by knocking down trees, which can crash into power lines and cause blackouts. High winds can have the same devastating effect, swinging power lines violently enough to yank them off the poles. When tornadoes rip through an area, they can tear power lines and poles apart. It may take weeks for utility companies to replace the lines and restore power.

Utility companies pay close attention to severe weather conditions and issue alerts to warn customers of potential power outages. They also take preventive measures to keep power running smoothly. They often have workers who monitor and trim trees that might fall against electrical wire during a storm. In addition, utility workers often check power lines and poles regularly to be sure they remain in good condition.

Lightning strikes can destroy equipment ▶
and cause blackouts over large areas.

One State's Crisis

Throughout 2001, residents of California were stuck in a major power crisis. Some homes and businesses went for days without power. Utility companies had to order rolling blackouts in many areas.

One problem was that the state's demand for electricity had grown dramatically. From 1996 to 2001, demand grew at about 6 percent a year. Experts say the growth of technology-related companies in the state had caused a strain on the utilities.

Many people think deregulation of California's utilities had a lot to do with the power outages. Deregulation is a plan that allows companies to compete for business. In this case, California's major utilities sold most of their power plants to companies called electricity wholesalers that would sell electricity back to the utilities. These private companies were under no legal obligation, however, to increase their production of electricity to meet demand, even though demand was growing fast. There was not enough energy to go around and the wholesalers charged the utility companies a lot of money to keep the supply up.

Weather was also a problem. California gets much of its energy from states in the Northwest that usually depend on hydroelectric power. A lack of snow and rain in the Northwest in 2000 and 2001 resulted in less power for the utility companies, which then had less power to sell to California.

To keep up with the energy demand, California's major utility companies had to buy energy from other companies and states, which was an expensive short-term plan. Because laws prevent consumer energy prices in California from rapidly increasing, the electric companies could not charge customers a higher price for the most expensive outside energy. As a result, the companies went into debt and could no longer afford to buy power from outside power companies. At that point, the federal government stepped in and required the companies to sell power to California.

Flick a Switch

Excited Electrons Create Electrical Current

Insulated Wires Pass Through Junction Box Behind Switch Plate

Color-Coded Wires Perform Different Jobs

Electricity Flows to Junction Box Wires from Circuit Breaker Panel

Service Head Connects Outside Wires to Home

Outside Home, Wires Connect to Utility Pole's Transformer

Electricity Travels Along Distribution Line to Utility Pole

How does electricity get to the pole and transformer drum?

It travels along a distribution line. This is a wire that runs from one utility pole to the next. The distribution line delivers electricity to every transformer box it passes.

The electricity that travels along the distribution line comes from the distribution bus. Although it has no wheels, this bus helps electricity travel. The distribution bus is a steel structure with power lines running from it. It splits the power in multiple directions so the power can travel to homes in different areas. The power from the distribution bus travels along the power lines that head out to electrical poles via a distribution line. Before electricity can travel on the distribution line, its voltage must be reduced.

What reduces the voltage?

Distribution lines ▶ can snap in severe weather, interrupting the power supply

▲ *Distribution lines carry electricity to neighborhood transformers.*

Earth Power!

All the outlets and switches in a home are connected to grounding wire, which carries electricity safely to the ground. It turns out that the ground conducts electricity, too. Earth is electrically charged.

Scientist Nikola Tesla (1856–1943) discovered the electrical power of the earth. He found that the planet acted like a giant metal ball. Tesla sent an electric current through the earth at his Colorado Springs laboratory. That electric current traveled a great distance and then returned to its point of origin. Each time the current came back to him, Tesla strengthened it with another electrical blast. He found that the electrical current traveled through the earth easily, because the planet's core is made of iron. Iron, a metal, is an excellent conductor of electricity. Tesla believed his findings could provide electricity to everyone around the world, free of charge. However, Tesla lost his financial backing to finish his electricity experiments. Since then, others have tried harnessing the power of the earth, but none have succeeded in supplying power to homes or businesses.

Flick a Switch

Excited Electrons Create Electrical Current

Insulated Wires Pass Through Junction Box Behind Switch Plate

Color-Coded Wires Perform Different Jobs

Electricity Flows to Junction Box Wires from Circuit Breaker Panel

Service Head Connects Outside Wires to Home

Outside Home, Wires Connect to Utility Pole's Transformer

Power Substation Reduces Voltage to Distribution Lines

What reduces the voltage?

The power substation—a large outdoor area with power towers, wires, and transformers—has the important job of stepping down, or transforming, high voltages of electricity to lower voltages suitable for travel through distribution lines. Huge transformers at the power substation must step down voltages in the hundreds of thousands to only about 10,000 volts. When the electricity travels to the power substation from the transmission substation, it flows through heavy-duty high-voltage transmission lines. The power lines near homes cannot carry as much electricity as a plant can send out, so the transformer has to lower the voltage.

Just like the circuit breaker panel in a house, the power substation has huge switches. These can be turned off to disconnect the substation from the distribution lines or the transmission substation whenever necessary.

How does the electricity get to the power substation?

Some employees of power ▶ substations must work with many thousands of volts of electricity.

Electricity Travels Along Distribution Line to Utility Pole

Power Substation Reduces Voltage to Distribution Lines

Transmission Substation Sends Electricity to Power Substation

Power from Plants Combines at Transmission Substation

Plant Sends Power to Nearby Transmission Substation

Plant's Turbine Turns Rotor to Create Electricity

▲ *Power substations bring voltage down to safe levels.*

Plan Ahead!

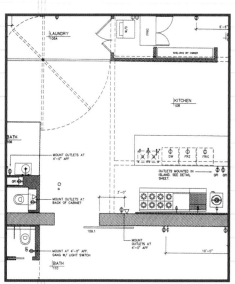

▲ *This blueprint shows where electrical outlets will go.*

Before a home is built, electrical contractors survey each room plan to figure out the best place for switches and outlets. Using the blueprints, or home's layout plan, electricians decide how much electrical wiring is needed, where ground wires go, which switches need to carry 240 volts for larger appliances, and how they will hook up enough wiring through the home.

Houses built before the 1960s usually had about 100 amps of power. Today, people have many more appliances than they did even five years ago. Computers, fax machines, modems, additional telephones, and Internet cable lines require more electricity than some homes can handle.

To upgrade older homes, electrical contractors assess a home's electrical needs and provide the correct amps to power up all the latest gadgets. They add on enough electrical switches, outlets, and wiring to the rooms that require it most. Usually, they will also install a new circuit breaker panel that can provide 150 amps or more.

Transmission Substation Sends Electricity to Power Substation

How does electricity get to the power substation?

A transmission substation sends electricity to the power substation. Like the power substation, the transmission substation has large transformers to change electricity. At this substation, however, electricity is stepped up, or increased.

Energy from a power generator enters the transformers at a few thousand volts. This is not nearly powerful enough to send electricity to the power substation, which can be hundreds of miles away. For this reason, the transformer converts the low voltage into extremely high voltages—up to 765,000 volts—so that it can travel the long distance to the power substation.

Towers and power lines that lead from a transmission substation are different from the power lines near homes. These heavy-duty, thick wires run through enormous steel towers, which are usually placed several hundred feet apart. The towers can continue in a line for miles on end until they reach the power substation. Because these lines can be so far apart, power can be lost along the way. The utility company regularly monitors voltages flowing to its substations, and can add power when necessary.

How does the utility company add power?

▲ *Wires supported by huge steel towers carry power away from a transmission substation.*

Electricity Travels Along Distribution Line to Utility Pole

Power Substation Reduces Voltage to Distribution Lines

Transmission Substation Sends Electricity to Power Substation

Power from Plants Combines at Transmission Substation

Plant Sends Power to Nearby Transmission Substation

Plant's Turbine Turns Rotor to Create Electricity

▲ *Transmission substations raise the voltage so the electricity can reach its destination.*

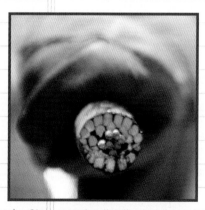

▲ *Close-up of a distribution line*

Alternating Current

The electricity that is sent from the power company to homes and businesses travels in the form of alternating current. Unlike the direct current of a battery, in which the electrons flow in only one direction, the electrons in alternating current switch back and forth constantly. Alternating current can be transformed to lower and higher voltages, while direct current cannot. Therefore, the power company uses this form of electricity to transmit and distribute power to homes and businesses. In a typical household outlet, the electrons in an outlet reverse directions fifty or sixty times every second.

Power from Plants Combines at Transmission Substation

How does the utility company add power?

Utility companies often buy power from power manufacturing companies, including other utility companies and independent power plants, and they combine that power with their own at the transmission substation. In other words, the power that people use to flick the switch and turn on the light may not come from their own utility company or even their own state. Once combined, the electricity continues to flow through the high-voltage transmission lines. Of course, all that power had to be produced first.

Where does all this power come from?

Power plants like the one at right, ▶ *owned by a Canadian power company in Quebec, may supply electricity to states hundreds of miles away.*

Can You Dig It?

Before homeowners drive a shovel deep into the ground near their home, they should call their local power supply company. If they are digging for home extensions, in-ground pools, and other major projects, they are required by law to call the power company first. This is because the ground around a home may be filled with buried utility wires, pipes, and cables that could cause major problems to the local utility system if damaged. When homeowners call the utility to report that they would like to dig, the company will send a worker to their home to mark off areas where utility cables lie underground.

Plant Sends Power to Nearby Transmission Substation

Where does all this power come from?

Electrical power is created at a power plant and fed directly to the transmission substation on its property. Although power plants in the United States come in many varieties, more than 52 percent of the country's total electricity usage is powcred by coal.

Coal or any other fuel can be burned in a large furnace at a power plant. The energy it gives off as it burns heats water to make steam. That steam is used to produce electricity.

But how does the plant create electricity?

Wonder of Water

Hydroelectric plants use the force of moving water, not steam, to create electricity. The Hoover Dam outside Las Vegas, Nevada, supplies power for a wide area through hydroelectricity. Every second, enough water to fill fifteen swimming pools rushes from the Colorado River through the gigantic structure, where its power is harnessed to make energy. The Hoover Dam is sometimes called the "eighth wonder of the world" because of its spectacular size and tremendous power. It is 736 feet (224 m) high, 1,244 feet (379 m) wide at its crest, and 660 feet (201 m) thick. It contains enough concrete to pave a highway across the Atlantic Ocean. The water the dam holds back forms Lake Mead, which at 120 miles (93 km) is the largest man-made lake in the United States.

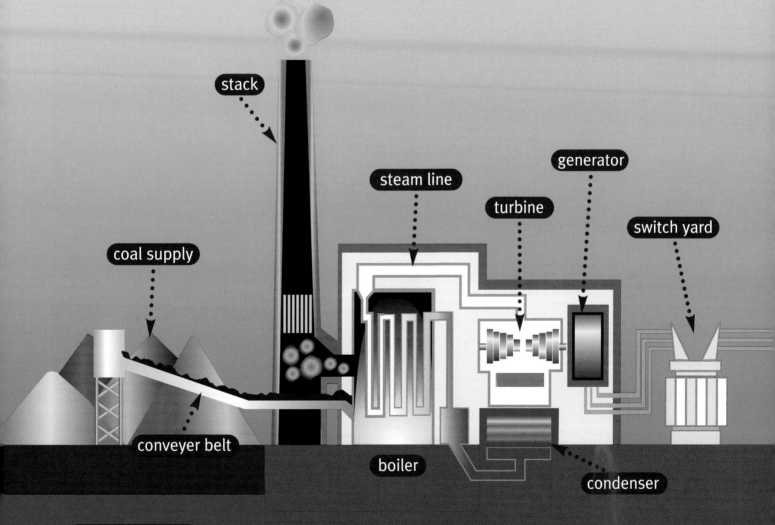

Electricity Travels Along Distribution Line to Utility Pole

Power Substation Reduces Voltage to Distribution Lines

Transmission Substation Sends Electricity to Power Substation

Power from Plants Combines at Transmission Substation

Plant Sends Power to Nearby Transmission Substation

Plant's Turbine Turns Rotor to Create Electricity

stack

steam line

generator

turbine

switch yard

coal supply

conveyer belt

boiler

condenser

river or reservoir

cooling water

The diagram above shows a typical coal-fired power plant

Flick a Switch

Excited Electrons Create Electrical Current

Insulated Wires Pass Through Junction Box Behind Switch Plate

Color-Coded Wires Perform Different Jobs

Electricity Flows to Junction Box Wires from Circuit Breaker Panel

Service Head Connects Outside Wires to Home

Outside Home, Wires Connect to Utility Pole's Transformer

Plant's Turbine Turns Rotor to Create Electricity

How does the plant create electricity?

When coal or other fuel is burned to heat water to produce steam, the high-powered steam hits a fanlike turbine and rapidly turns its blades.

When the turbine's blades spin, they turn the shaft. At the other end of the shaft is a part of the generator called the rotor. The rotor is made up of huge magnets inside copper wire coils. These magnets rotate when the shaft turns. The energy between the coils and the magnets creates an electrical current. (See the diagram on the facing page for an inside look at how a plant works.)

Generators make electricity in a ▶ process that begins with burning fuel.

| Electricity Travels Along Distribution Line to Utility Pole | Power Substation Reduces Voltage to Distribution Lines | Transmission Substation Sends Electricity to Power Substation | Power from Plants Combines at Transmission Substation | Plant Sends Power to Nearby Transmission Substation | Plant's Turbine Turns Rotor to Create Electricity |

Know Nukes

Nuclear power plants are becoming common for electricity production. They do not rely on fossil fuels, such as coal or oil used in traditional power plants, to make electricity. Because fossil fuels will eventually run out if they are used up too fast, the number of nuclear power plants is increasing each year. They provide about 17 percent of the world's electricity. In the United States, more than 104 nuclear power plants are in operation.

Nuclear power plants and hydropower plants both use steam as a force to spin magnets to create electricity. The major difference between the two kinds of plants is what is used to heat the water and make steam.

Nuclear power plants use nuclear reactions as their heat source to create electricity. This process is called fission. Fission causes the nucleus, or center, of an atom to split. Nuclear power plants split the nuclei of atoms of uranium, an element found naturally on Earth. That

▲ Nuclear reactors use uranium to start the process of making electricity.

splitting gives off heat and triggers the splitting of more uranium atoms in a nuclear chain reaction. This chain reaction produces large quantities of heat, which is used to turn water into steam. The rest of the process is like a coal-fired power plant: Steam drives a turbine that spins a generator to make electricity.

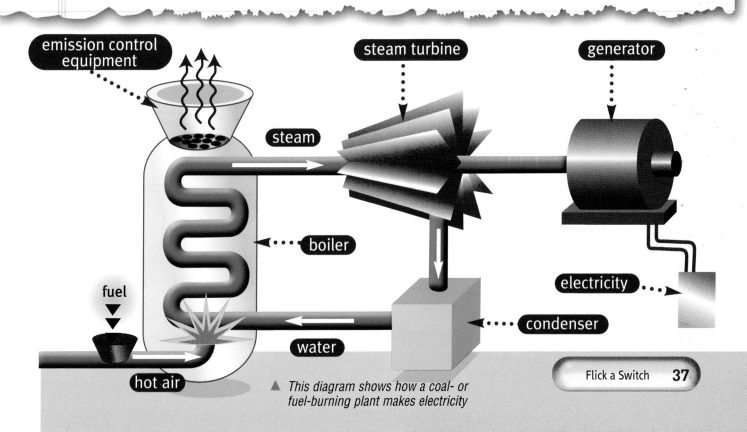

▲ This diagram shows how a coal- or fuel-burning plant makes electricity

ENERGY-PACKED!

How else can power be created?

Although the United States uses mostly coal-powered plants, other methods and materials can produce electricity. All of these methods generate electricity the same way: They use energy to turn a turbine. They only differ in the kind of material used to generate that energy.

Petroleum oil and coal are fossil fuels used in many power plants. These fuels are natural materials found in the earth. When burned in huge furnaces, they can heat water and make steam to turn a turbine.

Natural gas is another energy alternative. The gas is combustible, which means burning it can heat water for steam. It can also generate electricity without using water. Burning natural gas produces other extremely hot gases, which shoot through a turbine to spin its blades and generate electricity.

One of the cleanest energy makers is the sun. Light from the sun can be absorbed by a solar cell, which is made of materials that transform light into electricity. The main problem with solar power is that it is not available all day or even every day. That is why solar power makes up less than one percent of the electricity produced in the United States.

Wind is another natural resource used to generate electricity. Wind-turbine towers, which look like slimmed-down windmills, use energy in wind to produce electricity. Like solar power, wind power is not always available. Although wind turbines are increasingly popular in western states, such as Arizona, Nevada, and California, wind power accounts for less than one percent of electricity in the United States.

Another option, geothermal power, uses extreme heat from beneath the surface of the earth. The

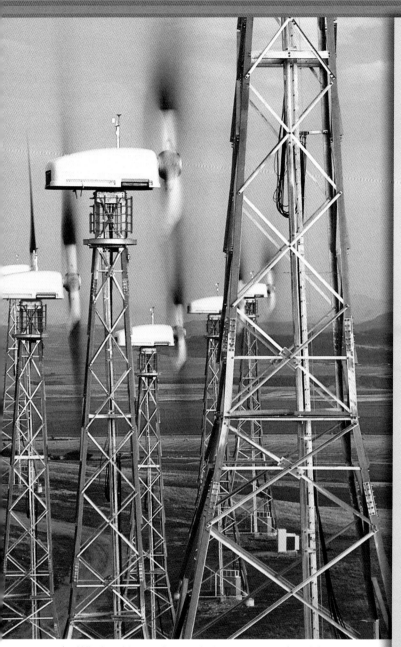

▲ *Wind turbines rely on wind to generate electricity.*

Spaced Out

Astronauts working in space need electricity to power their tools and light their capsules. Because it is impossible to hook up electrical poles and wiring from Earth to space, they use solar cells to convert light from the sun directly into electricity.

The International Space Station will use solar cells in the same way, except on a much bigger scale. The station looks like a giant dragonfly with extra-long wings. It has four sets of giant solar wings, each pair measuring 240 feet (73 m) long. Altogether, the wings contain more than 250,000 solar cells, which enable them to absorb enough solar power to light up an entire neighborhood.

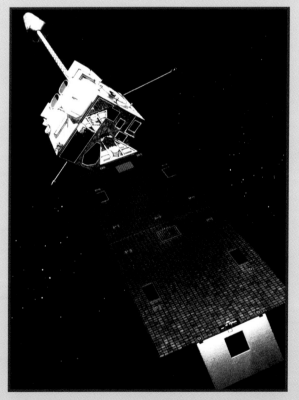

▲ *Panels of solar cells generate electricity when light shines on them.*

heat comes from volcanic activity deep below the ground. The heat mixes with underground water to make steam.

Yet another alternative is biomass, which consists of wood, solid waste, and agricultural waste. This material is burned to heat water for steam. Biomass generates less than one percent of electricity in the United States.

The Big Picture

13 Flick a Switch

Flick on a switch to turn on a light.
(pages 6–7)

11 Insulated Wires Pass Through Junction Box Behind Switch Plate

Electrical wires are housed in a junction box to keep current contained.
(pages 10–11)

9 Electricity Flows to Junction Box Wires from Circuit Panel

Electricity gets to the box from the circuit breaker panel.
(pages 14–15)

7 Outside Home, Wires Connect to Utility Pole's Transformer

Electricity passes through electrical wires that are secured high atop utility poles. There, a transformer converts the high-voltage electricity to a usable electrical current.
(pages 20–21))

12 Excited Electrons Create Electrical Current

Wires behind the switch plate carry charged particles that bring electricity to the light.
(pages 8–9)

10 Color-Coded Wires Perform Different Jobs

Current is carried to and from a light by different wires. Another wire provides an emergency exit for excess electricity.
(pages 12–13)

8 Service Head Connects Outside Wires to Home

The cane-shaped service head near the roof holds service wires from outside the home, allowing electricity to enter.
(pages 16–17)

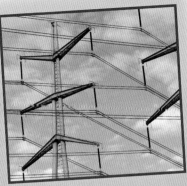

6 Electricity Travels Along Distribution Line to Utility Pole

Electricity is distributed to the wires via distribution lines and split up by distribution bus so it can reach different neighborhoods.

(pages 26–27)

4 Transmission Substation Sends Electricity to Power Substation

In order for electricity to travel the long distance to the power substation, its voltage is stepped up at the transmission station.

(pages 30–31)

2 Plant Sends Power to Nearby Transmission Substation

At a typical power plant, electricity production starts with burning coal.

(pages 34–35)

5 Power Substation Reduces Voltage to Distribution Lines

Power substation steps down voltage so it can travel on distribution lines.

(pages 28–29)

3 Power from Plants Combines at Transmission Substation

The local utility company sometimes gets power from other companies and independent power plants, and combines it at the transmission substation.

(pages 32–33)

1 Plant's Turbine Turns Rotor to Create Electricity

Turbine's blades are powered by steam to turn the rotor's magnets and create electricity.

(pages 36–37)

Facts and Figures

Over the years, there have been many key events in the development and uses of electricity. Here are a few:

1600

Static electricity is discovered when the physician to Queen Elizabeth I rubs a piece of amber on wool and creates an electric charge.

1751

Benjamin Franklin performs an experiment to show that iron and other metals are electricity conductors.

A painting showing Franklin's experiment

1752

Franklin's famous kite experiment proves that there is electricity in Earth's atmosphere. In addition, he invents the lightning rod.

An early battery ▶

1800

Italian scientist Alessandro Volta experiments with a battery-like structure using metal coil and chemicals and finds a way to produce an electrical current.

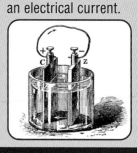

1878

In England, physicist Joseph Swan unveils the first practical lightbulb that uses a carbon filament to glow.

Lightbulb ▶ with carbon filament

1879

Inventor Werner von Siemens demonstrates the first electrically powered locomotive.

Thomas Edison improves on Joseph Swan's lightbulb. A year later, electric streetlights illuminate New York City.

1881

To provide electricity to city dwellers, Edison creates the Edison Electrical Illuminating Company, the first electric company in the United States.

The first ▶ steam turbine

1884

Charles Parsons creates the first steam turbine that is practical enough for consistent use. His design is still the basis for turbines used to produce power.

Watt a Lot!

- Large power plants produce about a million kilowatts of power each hour. That is enough to meet the power needs of about one million homes for an hour.

- Average American homes use about 24 kilowatt-hours of energy per day. Kilowatt-hours measure electricity, similar to the way inches are used to measure distance. One kilowatt-hour is a kilowatt of electricity used in one hour. In 24 hours, an average home uses 24 kilowatts of electricity.

- It takes about 714 pounds (324 kg) of coal to power a 100-watt light bulb 24 hours a day for an entire year.

- A stroke of lightning can produce about 100 million volts of electricity.

1820

French physicist André-Marie Ampère discovers that two current-carrying parallel wires are attracted to each other if the currents flow in the same direction. Later that year, German inventor Johan Schweigger uses Ampère's finding to make a machine that measures the direction and intensity of electric current.

1826

Working with Volta and Ampère's results, Georg Simon Ohm discovers that when there is a current in a circuit, there is also some heat and resistance to the current.

▲ Georg Simon Ohm

1832–1838

▲ Electric telegraph machine

American inventor Samuel Morse develops an electric telegraph machine.

1859–1873

Various inventors around the world engineer batteries and electrical generators, paving the way for electrically powered machines.

1888

Nikola Tesla invents the first practical alternating current electric motor, which runs on the same current used in our homes today.

1907

The carbon filaments previously used in lightbulbs give way to efficient tungsten filaments, still used today.

1954

Solar batteries that convert light energy into electric energy are developed by Bell Laboratories in the United States.

◀ Lightbulb with tungsten filament

1954

The first nuclear power plant begins operation near Moscow, in the Soviet Union, generating 5 megawatts of electricity.

Wonders and Words

Explore some common questions and misconceptions about electricity.

Q: *Why do some electrical outlets have three holes, and others have only two?*

A: The top two holes of an electrical outlet may look the same, but inside they are very different. The right slot is usually connected to a hot wire, a black or red wire that carries electrical current. The hot wire carries enough electricity to run the appliance. The left hole is usually connected to a neutral, or white, wire. It returns electricity at zero voltage. Some outlets have a third round hole on the top or bottom of the outlet. That hole is for the ground wire. It can be dangerous to use a hole adapter to plug a three-hole plug into a two-hole outlet, because this disables the ground wire. In the event of a short circuit, the electricity will have nowhere to flow and might start a fire.

Q: *What could happen if you plug in an appliance, such a hair dryer, over a bathtub or a sink?*

A: If an appliance falls into a sink or tub filled with water, the water will cause a short circuit and electricity will flow into the water. Metal plumbing will conduct that electricity and start up an enormous electrical flow that is nearly impossible to stop. If a person has contact with any part of the sink or tub, the electric current would flow through him or her and cause a potentially deadly shock.

Q: *Which gadgets use the most electricity? Which use the least?*

A:

Central Air Conditioner or Heat Pump	15,000 watts
Clothes Dryer	4,000
Water Heater	4,000
Water Pump	3,000
Refrigerator	1,500–1,900
TV	500
Computer	400
Dishwasher	350
Microwave Oven	300
Coffee Maker	106
Clothes Washer	103
Lightbulb	60
Blender	15

Q: *How can energy be conserved at home?*

A: In the winter, people can move their thermostat to 68 degrees or lower. For every degree the thermostat is lowered, a normal utility bill will be reduced by up to 3 percent. To keep heat from escaping or cold from coming in, windows and doors can be weatherproofed with insulation or weather strips. Homeowners may also lower the temperature on their water heater to 120 to 140 degrees, and switch to fluorescent lamps. These lights use 70 percent less energy than regular lightbulbs and last ten times as long.

Glossary

Amps: the number of electrons that flow past a specific point in a circuit in a given length of time. The rate at which electrical power flows to a light, tool, or appliance

Blackout: a total loss of power, usually due to weather or high energy demands

Circuit: the continuous loop of electrical current flowing along wires or cables

Current: the movement of electrons along a conductor, such as metal wires

Electron: a small particle that travels around an atom and has a negative charge

Generator: the machine in a power plant that generates electricity by rotating magnets inside copper coils to generate an electrical current

Grounding wire: a wire (usually made of bare copper) used in an electrical circuit to conduct current to the earth in case of short circuit

Hot wire: a wire that carries electrical current, usually covered with black or red plastic or rubber

Hydropower plant: a place where large amounts of water are used to create steam to make electricity

Kilowatt-hour: one kilowatt of electricity used for one hour

Neutral wire: a wire (usually covered with white plastic or rubber) that returns electrical current at zero voltage to the source of electrical power

Rotor: a cylinder inside a generator that holds a series of large electromagnets

Transformer: a device that lowers electricity to the correct voltage for distribution through neighborhoods. A small, drumlike transformer on utility poles outside a home further reduces voltage so it can be used in a home.

Transmission substation: a place where power from a power plant is converted to high voltages for long-distance transmission

Turbine: a machine with fanlike blades that is connected to the generators at a power plant. As steam drives the turbine, the turbine's blades turn a shaft that spins the generator.

Service panel: a gray metal box near the place where electrical power enters a house, containing circuit breakers or fuses to protect each circuit in the home

Short circuit: an accidental contact between two wires that causes a circuit to shut off or a fuse to blow

Switch: a small handlelike device used to turn lights and appliances on and off

Volts: the strength or pressure of an electric current

Watts: the total electrical energy that is used. Wattage can be calculated by multiplying the volts times the amps. One thousand watts equals one kilowatt.

Index

Credits:

Produced by: J.A. Ball Associates, Inc.
Jacqueline Ball, Justine Ciovacco,
Andrew Willett
Daniel H. Franck, Ph.D., Science Consultant

Art Direction, Design, and Production:
designlabnyc
Todd Cooper, Sonia Gauba

Writer: Anne Prokos

Cover: Brooke Fasani: hand flicking light switch; Bud Russell/TheMeterGuy.com: service head; Photospin: transmission wires; Ablestock/Hemera: power plant; PhotoDisc, Inc.: generator.

Interior: Ablestock/Hemera: p.3, p.48 light bulb (background), p.23 Times Square, pp.24–25 lightning, pp.44–45 power substation; p.33 power plant, p.34 Hoover dam, p.37 nuclear power plant, p.39 wind turbines, satellite, pp.40–41 New York City skyline (background); Brooke Fasani: pp.6-7 hand flicking light switch, p.15 circuit breaker; p.43 boy holding lightbulbs; Library of Congress: p.8 oil lamp; p.27 Nikola Tesla, p.42 Benjamin Franklin, light bulb, early battery, steam turbine, p.43 Georg Simon Ohm, telegraph, light bulb; Sonia Gauba: p.9 diagram of electrical flow, p.10 diagram of electrical socket, p.11 diagram of copper atom, p.35 diagram of power plant, p.37 diagram of electricity production; John Garbarini Photography: p.11 junction box, p.13 open junction box; Photospin: p.11 copper wires, p.12 electrical socket, p.27 distribution lines, p.33 man digging, p.42 match stick; Bud Russell/TheMeterGuy.com: p17 service head (top and bottom); Todd Cooper: p.17 batteries; Georgia Transmission Corporation: p.19 electric meter, p.20 transformers, p.28 welder, p.29 power substation, p.30 transmission substation, p.31 transmission substation, close-up of wire; PhotoDisc, Inc.: p.21 transformers on pole, p.22 surge protector, p.36 generator; Courtesy of The Bakken Library and Museum, Minneapolis, MN: p.21 lightning rod

For More Information

www.eia.doe.gov/kids/electricity.html
The U.S. government's official electricity website offers a wealth of information illustrated with pictures, diagrams, charts, and graphs.

DiSpezio, Michael Anthony. *Awesome Experiments in Electricity and Magnetism. New York: Sterling, 2000.*

Glover, David. *Batteries, Bulbs, and Wires. New York: Kingfisher, 2002.*

Nankivell-Aston, Sally and Jackson, Dorothy. *Science Experiments with Electricity. New York: Scholastic, 2000.*

Parker, Steven. *Eyewitness: Electricity. New York: DK, 1999.*

Peters, Celeste A. *Circuits, Shocks, and Lightning: The Science of Electricity. Milwaukee: Raintree, 2000.*

Sneider, Cary I.; Gould, Alan; Wentz, Budd. *The Magic of Electricity. Berkeley: University of California, 1999.*

Taylor, Helen and Sweet, Stephen. *A Lightning Bolt Is Hotter than the Sun: And Other Amazing Facts About Electricity. Brookfield, CT: Millbrook Press, 1998.*

Tocci, Salvatore. *Experiments with Electricity. New York: Scholastic, 2002.*

VanCleave, Janice Pratt. *Janice VanCleave's Electricity: Mind-Boggling Experiments You Can Turn into Science Fair Projects. New York: John Wiley & Sons, 1994.*